# ASTRONOMY

BOY SCOUTS OF AMERICA
IRVING, TEXAS

**1985 Printing of the
1983 Revision**

# Requirements

1. Do the following:
   a. Sketch the face of the moon, indicating on it the locations of at least five seas and five craters.
   b. Within a single week sketch the position of the moon in the sky at the same hour on three different evenings. Explain the changes observed.
   c. Tell what factors keep the moon in orbit around the Earth.
2. Do ONE of the following:
   a. Photograph or locate on a map of the sky a planet at approximately weekly intervals at the same time of night for at least 4 weeks. Explain any changes noticed on the photographs or map.
   b. Find out when each of the five visible planets will be observable in the evening sky during the next 12 months and compile this information in the form of a chart or table.
3. Do ONE of the following:
   a. In a sketch show the position of Venus, Mars, or Jupiter in the sky at approximately weekly intervals at the same time for at least 4 weeks.
   b. Using a compass, record the direction to the sun at sunset at approximately weekly intervals for at least 4 weeks in spring or fall (for 6 to 8 weeks in summer or winter) and relate this information to the seasons of the Earth.
   c. With the aid of diagrams explain the relative positions of sun, Earth, and moon at the time of lunar and solar eclipses and at the times of new, first quarter, full, and last quarter phases of the moon.
4. Using the shadow of a vertical pole in sunshine, lay out a true north-

Copyright 1971
Boy Scouts of America
Irving, Texas
ISBN 0-8395-3303-9
No. 3303    Printed in U.S.A.    12M585

south line (a meridian). Then, using the line and the pole on another day, measure the altitude of the noon-time sun and determine your latitude.

5. Identify in the sky at least 10 constellations, four of which are in the zodiac. Identify at least eight conspicuous stars, five of which are of first magnitude. Then do the following:

   a. Show in a sketch the position of the Big Dipper and its relation to the North Star and the horizon early some evening and again 6 hours later the same night. Record the date and time of making each sketch.

   b. Explain what we see when we look at the Milky Way.

6. With the aid of diagrams (or real telescopes if available) explain the difference between reflecting and refracting telescopes. Describe the basic purpose of a telescope, and list at least three other instruments used with telescopes.

7. Do the following:

   a. Describe the composition of the sun, its relationship to other stars, and some effects of its radiation on the Earth's weather. Define sunspots and describe some of the effects they may have on this radiation.

   b. Identify at least one star that is red, one that is blue, and one that is yellow, and explain the meaning of these colors.

8. Do ONE of the following:

   a. Visit a planetarium or observatory and submit a report to your counselor both on the activities occuring there and on the exhibits of instruments and other astronomical objects you observed.

   b. Spend at least 3 hours observing celestial objects through a telescope or field glass, and write a report for your counselor on what you observed.

9. Name different career opportunities in astronomy. Explain how to prepare for one of them. List the high school courses most useful in beginning such preparation.

# Contents

# Reach for the Stars

*That's one small step for a man; one giant leap for mankind.*—Eagle
Scout Neil A. Armstrong, upon becoming the first man to set foot on the
moon, July 20, 1969.

In beginning work on the Astronomy merit badge, you are joining a
company of people, who, awed by the heavens, wondered about the
stars and studied the movements of celestial bodies. In your own way,
you are joining the company of Neil Armstrong. He was an astronaut,
not an astronomer, but like astronomers dedicated to advancing
humankind's knowledge of the universe.

You probably know something about astronomy already. Just to make
sure you're on the right wavelength before you plunge in, let's review
the fundamentals.

- The Earth is a planet and it moves around the sun, which is an
  ordinary star.
- There are eight other planets orbiting the sun. Some of the planets
  have satellites (moons) revolving around them. The whole group of
  sun, planets, and satellites is called the *solar system.*
- Our moon is a satellite and is the body nearest to us in space.
- The solar system is a small part of a large grouping of stars called
  the *Milky Way galaxy.* It contains about 150 billion stars.
- There are probably several billion other galaxies, each containing
  billions of stars. They are so far away that the light from many of
  them takes millions of years to reach the Earth.

With those fundamentals in mind, you're about ready to start. But
first check with your Scoutmaster to learn the name of your local
Astronomy merit badge counselor. Talk with the counselor about the
requirements and how you should meet them. He or she will be glad to
help you. Then begin your thrilling journey into the vastness of the
universe.

First stop—the moon.

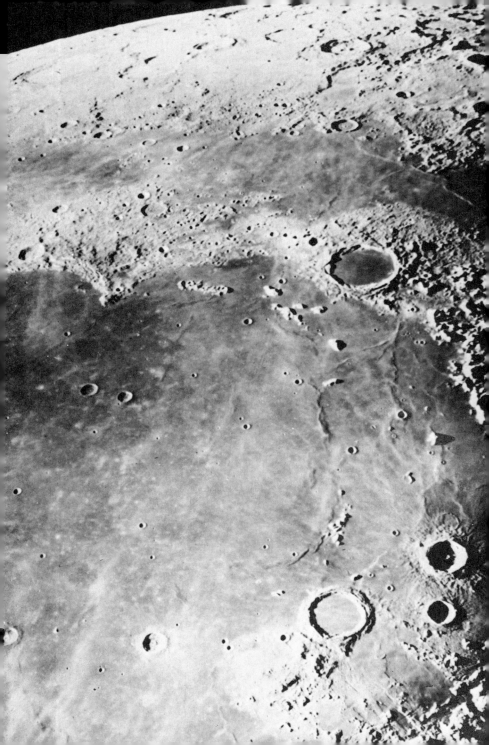

# The Moon—Our Nearest Neighbor

Because the moon is closest to Earth and was humankind's first target in its exploration beyond the Earth, we know a great deal about it. And because our astronauts were able to bring back samples of lunar rocks and soil and leave scientific instruments on the moon, our scientists are increasing their knowledge of the moon by leaps and bounds.

For this reason, we shall not discuss in detail the moon's composition, surface appearance, or probable origin. Your counselor will be able to direct you to the latest sources of information. Instead, we shall confine this discussion to the requirements.

## The Face of the Moon

You may have seen the surface of the moon on television, with astronauts lumbering about in the background in their bulky space suits. If not, you have seen still photos of the lunar surface.

You know, therefore, that it is not the smooth globe it appears to be to the naked eye. Rather, it is a drab, desolate landscape with craters and mountains and level plains called *seas*.

Some of you who read this will no doubt be our future astronauts and may visit the moon. On your return to Earth you'll probably say, "It's a nice place to visit, but I wouldn't want to live there." The moon has no discernible atmosphere, so there is no air to breathe. It has no water. During the lunar daytime (which lasts slightly more than 2 weeks) when the sun is shining on a spot on the lunar surface, the temperature is higher than the boiling point of water. During the lunar night (another 2 weeks) the temperature drops to nearly –300°F. There is no weather—no snow, rain, clouds, or fog.

*Mare Imbrium* (Sea of Rains). Plato, the large, dark crater near the top, is 60 miles in diameter.

# The Seas and Craters

At the next full moon, go out at night and take a good look at it. You will see a number of dark areas. These are called *seas* because the astronomers who trained the first crude telescopes on the moon in the 17th century thought the dark areas were seas or oceans. With a small telescope or a pair of binoculars, you can see how they arrived at that idea.

We know that there is no water on the moon and that the seas are actually broad, level plains. Most of these plains are ringed by mountains and pitted by craters of varying sizes.

The biggest of these plains earned the title of ocean. It is called *Oceanus Procellarum* (Ocean of Storms) and is about 900 miles (14,480 meters) across. Some of the smaller plain areas were named marshes, swamps, bays, and lakes.

There are more than 20 seas on the moon. Some of the more prominent:

*Mare Imbrium* (Sea of Rains)
*Mare Serenitatis* (Sea of Serenity)
*Mare Tranquilitatis* (Sea of Tranquility)
*Mare Vaporum* (Sea of Mists)
*Mare Frigoris* (Sea of Cold)
*Mare Nubium* (Sea of Clouds)
*Mare Nectaris* (Sea of Nectar)
*Mare Fecunditatis* (Sea of Fertility)
*Mare Crisium* (Sea of Crises)


The moon's surface is pockmarked by thousands of craters—bowl-shaped cavities or pits. Some are more than 100 miles (161 kilometers) wide, and one is at least 29,000 feet (8,800 m) deep. The broadest of the moon's craters often are called *walled plains* because their sides are much like steep cliffs rising out of the bottom to mountain height. Many astronomers classify the moon's craters according to their sizes and conformations, using names such as ringed plains, craterlet, cratercone, and craterpit.

For a long time scientists have debated the origin of the moon's craters. Some believed they were caused by volcanic activity within the moon. Others suggested that craters are scars left by meteors that have struck the moon for billions of years. Objections have been raised to both theories. Talk with your counselor to learn the current thinking of scientists about the origin of the moon's craters.

To map a few of the moon's seas and craters for requirement 1a, get a book about the moon from your school or public library. Many have fairly detailed maps showing the major seas and craters.

## The Moon's Travels

Now let us come back to Earth again and look at the moon from here. But before we look up, let's be sure we are clear about what the moon is doing.

First, we know that it is orbiting the Earth. Since we see it rise in the east and set in the west, it appears to be traveling westward. Actually, it is orbiting us from west to east, but because the Earth is rotating on its axis from west to east faster than the moon is traveling in orbit, we are *passing* the moon. You can check this by watching the moon's position against a star to the east of it. As the hours pass, you will see that the moon will draw closer to the star even though the moon seems to be moving westward.

On the average, the moon takes 27 days, 7 hours, 43 minutes, and 11.47 seconds to make one trip around the Earth. But this is not the actual time from new moon to new moon. Remember, the Earth is moving, traveling in its orbit around the sun. It takes the moon about 2¼ more days to return to its starting position relative to the sun. And so, on the average, the time for the moon to make the circuit from new moon to new moon is 29 days, 12 hours, 44 minutes, 2.78 seconds. The time may vary by as much as 13 hours. Ancient people probably used the phasing of the moon as their measure of time.

We know that the moon shines because it reflects sunlight. It has no light of its own. Although it often seems bright, it is really a poor reflector. Only 7 percent of the sun's light that strikes the moon is bounced off it. By contrast, the Earth is a fairly good reflector. Astronauts found that from the moon the Earth is very bright; it is estimated that about 40 percent of the sun's light is reflected by Earth.

During the 29-plus days that it takes the moon to complete its full circuit with respect to the sun, its appearance varies greatly. Just before and after new moon we see a slender crescent. At the first- and third-quarter phase we see the moon half-illuminated. At full moon we see a completely illuminated disk.

The moon, of course, has not changed; it is always a big ball, roughly

2,160 miles (3,575 km) across. Why, then, does it look different?

It looks different because while one-half of the moon is always lighted by the sun, we cannot always see all of the lighted side. The drawings on page 11 will make it clear why this is true.

When you have watched the moon for a few nights, you will notice that its position in the sky is slightly different each day at the same

**Moon's orbit around the Earth; arrow points to sun.**

**Phases of the moon**

11

hour. You will notice that it rises later each night. The reason is that while the Earth is making one full rotation to mark its day, the moon also is moving in its orbit. Each night finds the moon about 13.2 degrees farther to the east. Therefore, the Earth has to turn more than one revolution to "catch up" to the moon.

To make your sketch for requirement 1b, draw a couple of the more prominent constellations (see pages 43–54) on the first night of your viewing. Draw in the moon. Then, at the same hour on two other nights that week, go out again and draw in the moon's new position.

## The Moon's Orbit

The moon's orbit around the Earth is an ellipse, or slightly flattened circle.

At its *apogee* (when it's farthest from the Earth), it is 252,710 miles (406,610 km) away. At its *perigee* (closest to the Earth), it is 221,463 miles (346,334 km) away. The mean distance is 238,857 miles (384,321 km).

By traveling in an elliptical orbit, the moon is following one of the natural laws that govern the universe. All planetary bodies do the same thing. The Earth itself follows an elliptical orbit around the sun.

What keeps the moon spinning around the Earth? Two forces are at work on it. Sir Isaac Newton (1642–1727), one of the greatest scientists in history, explained them. The first is the law of motion, or law of inertia. It states that every body at rest tends to remain at rest and every body in motion tends to remain in motion unless acted upon by a force.

Astronomers don't know where the moon got its original motion. But we do know that it is traveling in its elliptical orbit around the Earth at an average speed of more than 2,200 miles an hour (3,540 km/hr). The moon travels faster when it is in perigee—near the Earth—and slower in apogee.

If you tie a ball on the end of a string and whirl it around your head, you know that if you let go the ball is going to fly off away from you. Why doesn't the moon do that?

The answer is the second force that is acting on it. That force is gravity. Newton, who explained gravitation as well as the law of inertia, said that every body in the universe attracts every other body. How strong the attraction is depends on two things: the masses of the two bodies and the distance they are apart.

The law of gravitation explains why a ball falls to earth when you throw it. And it explains why the moon goes around the Earth without sailing off into space. Because the moon is a mass (only about one-eightieth the mass of the Earth), the Earth holds it in orbit by the action of gravitational forces.

But Newton said that every body is attracted by every other body. Does the moon, therefore, exert gravitation on the Earth? It does indeed. The moon's gravity influences tides on the Earth. It exerts a pull on the Earth that causes the planet to waver slightly as the Earth-moon system revolves around the sun.

## The Moon's Rotation

The moon has one other principal motion besides revolution around the Earth and, with the Earth, revolution around the sun. Like the Earth, the moon rotates on its own axis.

The moon only rotates once in the 27-plus days it takes to make a circuit around the Earth. It is for this reason that only a little more than half of the moon's surface is ever visible from Earth.

To prove that the moon rotates try this experiment. Put a chair in the center of your room. That's the Earth. You are the moon. Now *face* one wall and start moving around the chair; keep facing the *same* wall. You can see that if you, as the moon, keep facing in the same direction, then *all* sides of you will be visible at the same time during your revolution around the Earth. Now try moving around the chair *facing* it at all times. You will find that you have to face each wall in the room in turn; therefore, you have rotated once.

It would seem from this experiment that we should never be able to see more than half of the moon's surface from the earth. In fact, during the course of a month we can see almost 60 percent of the surface. This is due to a slight swinging of the moon caused by its irregular speed around the Earth and to the fact that its equator is tilted at a small angle to its orbit around the Earth.

PLUTO

NEPTUNE

URANUS

SATURN

JUPITER

SUN

MARS

EARTH

VENUS

MERCU

# The Planets

So far we have been talking about a heavenly body that is a long way off as we think about time and space here on Earth. Traveling at speeds of 4,000 to 24,300 miles an hour (6,435 to 39,100 km/hr), it still took our astronauts more than 2½ days to reach the moon.

But now we must expand our thinking to include the whole solar system, which covers much greater distances. As space is measured, the moon is really only an arm's length from the Earth. If astronauts took off from Earth for the planet Pluto and traveled the whole way at a brisk 25,000 miles an hour (40,225 km/hr), it would take them more than 12 years to reach that most distant planet in the solar system. So you can see that we must now broaden our horizons considerably.

As we have said before, the solar system is made up of an ordinary star (the sun), nine planets including the Earth, and our moon. It also includes at least 51 other satellites or moons, possibly more than 100,000 asteroids (minor planets) ranging in size from a few feet to a few hundred miles in diameter, and an unknown number of comets. Some estimates put the number of comets at 100 billion.

The king of this gigantic realm is the sun. We will put off a discussion of His Majesty for a while and discuss his principal subjects—the planets. Only five of them besides the Earth are visible to the naked eye. In perfect weather you may see a sixth—Uranus. The other two are at such vast distances that only powerful telescopes can give us a dim look at them. Like the moon, planets do not produce light, they merely reflect the light of the sun. Let's take a look at the vital statistics of the planets and then consider them one by one.

| | Average Distance From Sun in Miles | Period of Solar Circuit | Period of Axial Rotation | Mean Diameter in Miles |
|---|---|---|---|---|
| Mercury | 35,960,000 | 88 days | 59 days | 3,000 |
| Venus | 67,200,000 | 225 days | 243 days | 7,500 |
| Earth | 92,960,000 | 365.25 days | 23 hrs. 56 m. | 7,927 |
| Mars | 141,600,000 | 687 days | 24 hrs. 37 m. | 4,220 |
| Jupiter | 483,600,000 | 11.9 years | 9 hrs. 50 m. ± | 88,700 |
| Saturn | 886,000,000 | 29.5 years | 10 hrs. 39 m. ± | 75,100 |
| Uranus | 1,782,000,000 | 84 years | 10 hrs. 49 m. ± | 29,200 |
| Neptune | 2,801,000,000 | 164.8 years | 15 hrs. 40 m. ± | 27,960 |
| Pluto | 3,661,000,000 | 247.7 years | 6.4 days | 1,500 |

**Relative sizes in the solar system**

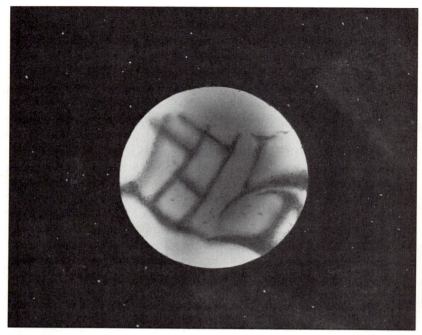

## Mercury

Mercury, the planet closest to the sun, is also the speediest. It zips through its orbit at speeds from 23 to 35 miles a second (37 to 56 km/sec.) and requires only 88 days to complete its circuit around the sun. Thus a year on Mercury is only 88 days long.

Mercury rotates on its own axis once every 59 Earth days with respect to the stars. A day on Mercury, in relation to the sun, is about three times as long. During the long daylight period on Mercury (about 90 of our days), the temperature rises to more than 700°F. During the long night, the temperature falls to less than –300°F. There is no sign of an atmosphere on this small planet.

## Venus

Of all the planets, Venus is most similar to the Earth in size. It is only about 427 miles (687 km) smaller in diameter, and its mass, density, and gravitational pull are all only a little less than the Earth's. It can also be the closest planet to the Earth and, except for the sun and moon, the brightest object in the sky.

Like the Earth, Venus has an atmosphere. In fact, it has such a thick, cloudy atmosphere that we can't see through it to look closely at the planet itself. The atmosphere is mostly carbon dioxide that presses on its surface about 100 times as much as on Earth. Venus has little of the

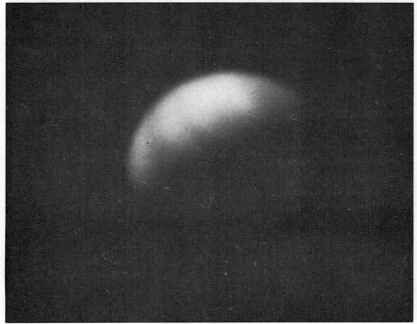

**Venus**

oxygen that supports living things on Earth. Temperature at the surface is nearly 1,000°F., and the clouds are made of battery acid ($H_2SO_4$).

Venus whips along its nearly circular orbit of the sun at an average speed of 22 miles a second (35 km/sec.), completing its year in 225 days. It rotates slowly "backward." A day (sunrise to sunrise) on Venus is about 116.8 Earth days, dark being one-half this period.

# Mars

Next outward from the sun in the family of planets is the Earth. Then comes Mars, perhaps the most interesting to astronomers because there is some chance that primitive forms of plant life did exist there. The Viking landers, however, found no positive evidence for life.

Mars is a small planet—a little more than half as big as the Earth. A day there is 24 hours, 37 minutes—very close to Earth's day—but a Martian year is almost two of Earth's years. Temperatures range from

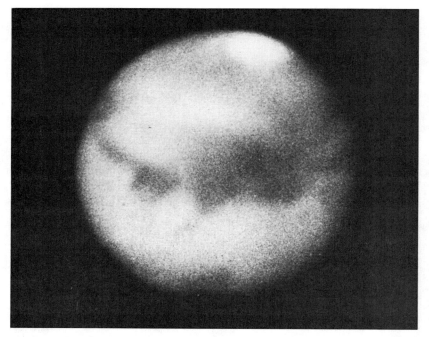

**Mars**

80°F. during the day at the equator to −200°F. at night near the poles. Nighttime temperatures are generally −150°F. because Mars has a thin atmosphere of carbon dioxide with little water vapor or oxygen. The planet has two small satellites, or moons.

## Jupiter

More than 300 million miles out from Mars is Jupiter, the giant of the planets. It contains more material than all the other planets combined, and its diameter is more than 11 times that of the Earth.

Jupiter has an extremely dense atmosphere. The planet itself is composed chiefly of light substances such as hydrogen, methane, and ammonia, which are gases on Earth.

**Jupiter—this huge planet has 17 moons**

Jupiter spins rapidly, so a day there is only 9 hours, 50 minutes long. One of its years is almost 12 Earth years. Because of its rapid rotation, Jupiter bulges at its equator like the bloated giant it is.

## Saturn

Four hundred million miles farther in space is Saturn, next largest of the planets. Like its big brother Jupiter, Saturn has a low density and bulges at its middle.

Saturn is best known for its rings, which are wheel-like disks less than 10 miles thick. *Voyagers I* and *II* revealed the rings to be very complex and full of surprises. The rings are made up of snow and grit and may be the debris of what was once a moon. Saturn also is orbited by 22

conventional moons, the largest of which is as big as the planet Mercury. Saturn's day averages 10 hours, 39 minutes long, but its year is almost 30 of our years.

## Uranus

Uranus, which occasionally is just visible to a person with good eyesight, is 900 million miles beyond Saturn. It is the third largest planet with a diameter of 29,200 miles (46,982 km). Its atmosphere is composed of hydrogen and methane. Because of its vast distance from the sun, it is, like Jupiter and Saturn, a forbiddingly cold place. Its temperature is −300°F. It was unknown until 1781.

Uranus rotates on its axis in 10 hours, 45 minutes and takes 84 years and 4 days to revolve around the sun. It has five known satellites. Arrows indicate three of them. Uranus has a faint system of rings.

## Neptune

A billion miles beyond Uranus is Neptune. It is slightly smaller than Uranus but resembles it in atmosphere and speed of rotation. Neptune

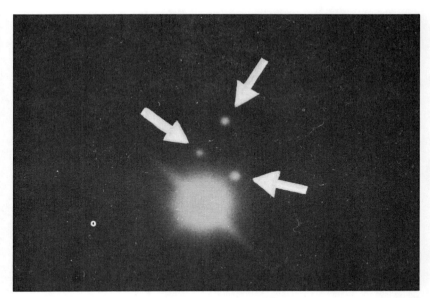

**Uranus**

**Neptune**

rotates in about 16 hours, but it takes almost 165 years to go around the sun.

Neptune was the first planet to be "discovered" by mathematics. Astronomers observing Uranus during the early part of the 19th century noticed irregularities about its orbit that could only be explained by the attraction of another planet. (The law of gravitation, remember?) Mathematicians calculated from Uranus' irregular movements where another planet should be. In 1846, an astronomer aimed his telescope at that portion of the sky and found Neptune within a half hour.

Neptune travels with three moons. One is indicated by the arrow.

## Pluto

As you might guess, Pluto also was searched for mathematically, but its discovery did not come until 1930. From December 11, 1978, to March 14, 1999, it will be 2.79 billion miles from the sun, closer to the sun than Neptune.

From Pluto the sun would appear about magnitude −18 (see page 55), far brighter than any star, but it would not show a visible disk. Pluto's day is believed to be about 6½ Earth days, and it takes nearly 250 years to circle the sun once.

Pluto, indicated by the arrow, is the smallest planet. It probably has no atmosphere, and its temperature is a frigid −350°F. Pluto's moon, Charon, was found in 1978, which enabled astronomers to "weigh" Pluto. It would take about 600 Plutos to equal the Earth. With a diameter of about 1,500 miles (2,400 km), Pluto's density works out to be about the same as water ice.

**Pluto**

# Asteroids, Comets, and Meteoroids

There are other members of the sun's family besides the planets and moons. The principal ones are asteroids, comets, and meteoroids.

## Asteroids

These are minor—most of them very minor—planets of the sun. Estimates of their number range from 30,000 to more than 100,000. Some are only a few feet in diameter. The largest, named Ceres, is about 480 miles (772 km) across.

The orbits of about 3,000 asteroids are known with some certainty. Roughly 85 percent of them are in orbit in the huge expanse of space between Mars and Jupiter. A few sometimes are nearer the sun than Mars, and a few orbit beyond Jupiter.

## Comets

An unknown number of comets are in the solar system, traveling in odd orbits. They are lumps of frozen gases and various kinds of metallic and stony grit.

Comets have no tails in the far reaches of space. But when they make a rare appearance in the vicinity of the sun, solar energy makes the comet stream out a tail that may be millions of miles long. Those that are bright enough to be seen without a telescope are rare and usually unpredictable. There may be two or three visible every 10 or 15 years. When you hear of a comet being sighted, make an effort to see it. It is truly an amazing sight.

## Meteoroids, Meteors, and Meteorites

These are different aspects of the same inhabitants of the solar system. They start as meteoroids—pieces of stony or metallic materials in space. Some are microscopic and others are huge. The Earth's atmosphere probably collides with 80 to 100 million of them every day.

If a meteoroid is large enough and if the collision occurs at night, the result is a meteor, or shooting star. The meteoroid burns in the tremendous heat created by its collision with our atmosphere, and we see for a fraction of a second a streak of light across the sky.

| METEOR SHOWERS | |
| --- | --- |
| **SHOWER** | **DATE** |
| Quadrantid | January 3 |
| Lyrid | April 21 |
| Eta Aquarid | May 4 |
| Delta Aquarid | July 28 |
| Perseid | August 11 |
| Orionid | October 21 |
| Taurid, North | November 1 |
| Taurid, South | November 16 |
| Leonid | November 17 |
| Geminid | December 13 |

Each day, one or two of them survive the flaming fall to Earth without burning up completely. The result is a meteorite. Most meteorites are quite small, but the largest known one weighs at least 60 tons. It fell during prehistoric times in Africa. The Willamette Meteorite, shown below, was found in Oregon's Willamette Valley in 1902. It weighs 15½ tons (14 metric tons) and measures 9.8 feet (3 m) long. It is the largest meteorite ever found in the United States, and is believed to have come from the asteroid belt.

Meteor showers are fun to watch. Shower meteoroids are believed to have once been parts of comets, and continue to follow in an orbital path. The Earth passes different meteoroid clusters throughout the year. The important meteor showers are listed here. They are named for the constellations from which they appear to come.

# Charting and Photographing Visible Planets

Now that we have a passing acquaintance with the solar system, let's turn to requirements 2a and 2b. Whether you choose to photograph a planet or chart the visible planets, you must first find out where they are.

This is not as hard as it might seem. All the planets except Pluto travel within a narrow band called the *zodiac*. This narrow band is centered on the ecliptic, or path that the sun appears to follow among the stars.

Therefore, if you note the stars that "set" shortly after sunset in just about the same place as the sun, you will have no difficulty after a while in figuring out the path of the zodiac.

However, even if you have a pretty good idea about the location of the zodiac, you may still have trouble finding the planets. They wander about the sky, keeping within the zodiac, and so it's easy to miss them.

Your best bet is to check a sky map. Visit a newsstand or library and ask for a publication, such as *Sky and Telescope* magazine, that will have sky maps for the time of year you need. The *World Almanac* also will list the positions of the planets for a year or more. If you can't find such a book check with your counselor. Here are a few tips on recognizing the planets:

## Mercury

This small planet is hard to spot even though it is quite bright. It is visible without a telescope about a dozen times each year—either just after sunset or just before dawn. It is always in the general direction of the sun and never more than 30 degrees from it.

## Venus

After the sun and moon, Venus is the brightest object in the sky. Sometimes it can be seen in daylight. It is never more than 3 hours ahead of or behind the sun.

## Mars

Although Mars is called the Red Planet with good reason, it is not always easy to find because it varies in brightness. Check its position in an astronomical almanac.

### Jupiter

Jupiter trails only Venus among the planets for brightness. It is not difficult to find if you know where to look, but its apparent progress is so slow that you might mistake it for a star unless you watch its movement for several nights.

### Saturn

You won't see the characteristic rings with the naked eye, but Saturn is not hard to find because it is bright. Not hard, that is, if you know where to look. It, too, takes a long time to show movement, so it is important to consult an almanac before you begin scanning the heavens for it.

### Uranus

This distant planet is just barely visible to the naked eye under good viewing conditions, but because it moves so slowly, it is easily mistaken for a star. Through a telescope it appears to be pale green, but to the naked eye it looks like an average star.

# Photographing a Planet

Unless you are fortunate enough to have access to the special cameras used in astronomy, use a Polaroid camera and 3000 speed Polaroid film in your effort to photograph a planet. Place the camera on a stepladder or other solid stand. Try several exposures of between 10 and 30 seconds at maximum—wide open—lens aperture.

The reason a Polaroid camera works best is because it uses fast film. With an exposure of between 10 and 30 seconds this camera can capture enough of the weak light from the planets and stars to show on film.

Most other ordinary cameras use much slower film. You can use them to shoot the stars and planets, but you must leave the shutter open much longer—perhaps a minute or two. If you leave the shutter open for many minutes or even hours, you will get star trails, as shown on the next page, on your pictures because of the turning of the earth during the exposure. Try longer exposures to see what happens.

**Circumpolar star trails**

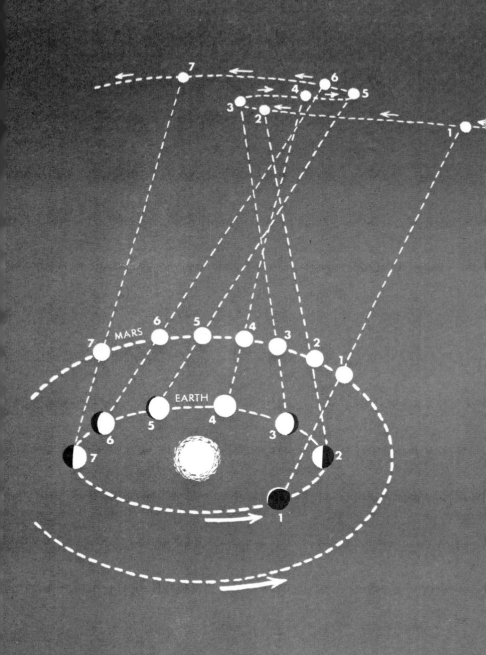

# Tracking a Planet

You can make a device to help you track a planet with a few simple items: a coat hanger, some pieces of cardboard (backs of writing tablets will do nicely), a sheet of clear plastic, and a marking pen. With these materials you can make a viewing device that will record accurately the movement of a planet against the background of a constellation or two.

Start by bending the coat hanger as shown on the next page. This will be the holder for your viewer.

Cut out rectangular centers from the cardboard sheets. These make the frame for your viewer.

Insert the clear plastic between the cardboard sheets. This is your viewer. When you mark the positions of stars and the planet you are tracking on the plastic, it becomes a star chart.

Put the frame with the plastic sheet into the loop in your holder. Place the wide end of the holder against your chin, holding the other end with your hand. Aim it toward the planet you are tracking.

Mark the positions of nearby stars and the planet with a marking pen while sighting through the viewer. *Hold it steady as you work.* Now you have the planet's position relative to the distant stars.

About a week later at the same time of night, line up your viewer *exactly* on the same stars. You will notice that the planet has moved in relation to those stars. Mark its position again on the plastic.

Do the same thing for 2 more weeks. The result will be a star chart showing the planet's movements. Transfer this chart to a sheet of paper, and you will have a pretty accurate sketch of the planet's course.

Keep your viewing device in a place where it won't become bent during this experiment. If it is bent after your first sighting so that the distance between your chin and the plastic sheet is changed, you may have trouble lining up the constellation the next time you want to sight one. Be sure to keep a record of the dates and times of your sightings.

In tracking planets, remember that they do not revolve around the sun at the same speed as Earth. Therefore, if you track Mars for half a year around its "opposition," you will find that from our viewpoint, it seems to reverse its field and travel "backward" for 3 months. But that's only because of the way we see it from Earth.

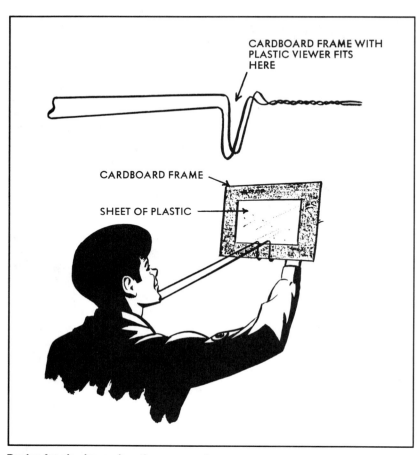

CARDBOARD FRAME WITH PLASTIC VIEWER FITS HERE

CARDBOARD FRAME

SHEET OF PLASTIC

**Device for viewing a planet's movements**

# The Sun and Our Seasons

As the Earth travels in its elliptical orbit around the sun our seasons change. During the Earth's annual orbit the direction to the sun also changes. Can you determine why you will have to record the direction to the sun for a longer time in summer or winter than in spring or fall to complete requirement 3b?

## Viewing the Sun

Never, never look straight at the sun through a telescope or binoculars, and never give it more than a passing glance with your naked eye. Serious damage and even permanent blindness can result from exposing your eyes to the direct rays of the sun.

If you want to look directly at the sun, you must filter out most of the light. There are several ways to do this. One way is to buy a cheap roll of *black and white* (not color) film, unwrap it to expose it, and have it developed. Cut up a few pieces about 5 or 6 inches long. Put two or more of them in front of your eyes as you look toward the sun. If you can't see anything, take one thickness away and look through it again. With only two thicknesses of film you should be able to see the sun clearly, even if the day is slightly overcast or hazy. A better way to view the sun safely is by projecting it through a telescope onto a screen as shown on page 77.

## The Reason for Our Seasons

Why do we have seasons? Why shouldn't the climate be the same during the whole 365¼ days it takes for the Earth to orbit the sun? Is it because the Earth is traveling in an elliptical orbit around the sun so that we're actually closer to it at certrain times of the year than others? No. Actually, we are slightly closer to the sun during winter than summer. (The mean distance to the sun is about 93 million miles; in July, we're 94½ million miles from the sun, while around the new year we're "only" 91½ million miles away.)

This doesn't seem to make sense, does it? Well, it will make sense

**Apparent path of the sun at different times of the year**

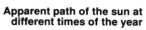

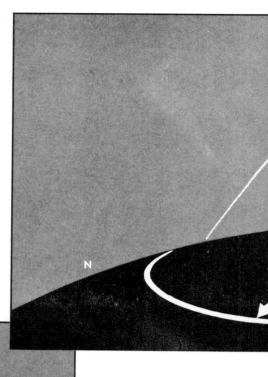

N

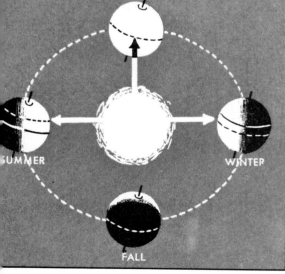

SPRING

SUMMER

WINTER

FALL

Tilt of the earth's axis relative to its orbit around the sun

when we understand that the Earth's axis is tilted slightly relative to our orbit around the sun.

Because the Earth's axis is slightly tilted, we receive the sun's rays at different angles at different times of the year. In the northern hemisphere in winter the rays come to us at a smaller angle than in summer; therefore, they spread themselves over a larger area. Also, we are exposed to the rays for a much shorter period each day during winter. Thus, our region receives and retains less heat from the sun.

From our viewpoint here on Earth, it seems that the sun rises in the east and sets in the west, but that is really true only on the first day of spring and the first day of fall.

In the winter, it rises southeast and sets southwest. During summer, it rises northeast and sets northwest.

You can check this by taking the sun's bearing at sunset once a week for at least 4 weeks in spring or fall, for 6 to 8 weeks in summer or winter. You will quickly see that its angle changes.

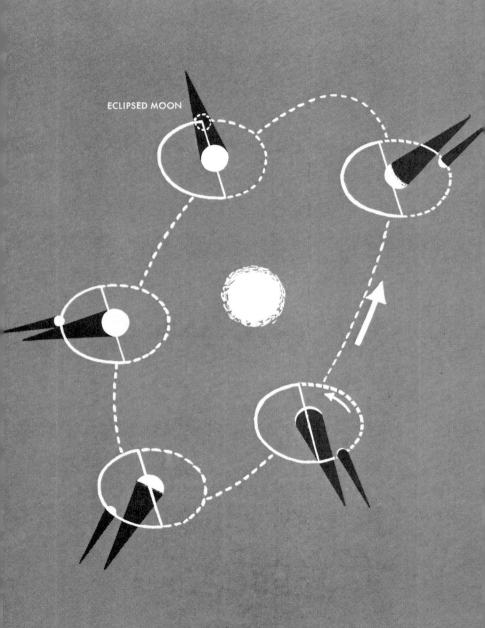

ECLIPSED MOON

# Eclipses and Phases of the Moon

An eclipse is a blackout of the sun or moon as seen from the Earth. When the new moon is exactly on a line between the sun and a section of the Earth, it is usually just big enough to blot out the sun's light. This is a solar eclipse.

When the full moon is within the shadow of the Earth we have a lunar eclipse. The moon is not quite fully blacked out; it still glows with a dim, copper-red light.

## Lunar Eclipse

Eclipses of the moon are not as spectacular as solar eclipses, but you have a lot better chance of seeing one. The moon is eclipsed when it moves into the Earth's shadow. At the distance of the moon the Earth's shadow is 5,690 miles (9,160 km) in diameter—more than twice the diameter of the moon. Therefore, the whole moon can easily enter this shadow, and when it does, it is seen in eclipse from over half the Earth.

When a lunar eclipse occurs, the moon is first dimmed by the Earth's penumbra, or partial shadow. Then it is slowly dimmed even further by the umbra, or darker shadow.

During each full moon, the Earth is between the moon and the sun. Why don't we have a lunar eclipse every month then? For the same reason we don't have a solar eclipse each month—because the moon's orbit is not in the same plane as the Earth's orbit around the sun.

## Solar Eclipse

There are two to five solar eclipses every year, but only a few of them are total; that is, when the moon fully darkens the sun. More often than not, only a part of the moon passes in front of the sun. Because the cone of shadow the moon casts is just long enough to

Orbits of the Earth and moon, showing shadows cast by each and lunar eclipse when Earth's shadow covers moon.

reach the Earth, a total eclipse is seen in a rather narrow band—never more than 170 miles (273 km) wide. An annular eclipse happens when the moon appears too small to cover the sun.

We don't have a solar eclipse every time there is a new moon because the moon's orbit is tilted about 5 degrees. At most new moons the moon does not line up directly between the sun and the Earth. Most of the time it passes either a little above or below the sun.

Your chances of seeing a total solar eclipse are not good. Before the year 2000 there will be only three solar eclipses visible from the United States:

May 30, 1984—southern United States (barely total).

July 11, 1991—Hawaii, better seen in Mexico.

May 10, 1994—Texas to Maine (annular eclipse).

There will be plenty of other total eclipses in other parts of the world during this century. If you become an astronomer, no doubt you will travel to see one, because during eclipses astronomers can find out things about the sun that are impossible at other times.

Even if you don't become an astronomer, you usually can find a tour to join that will let you experience all the excitement of eclipse watching.

# Phases of the Moon

We have already talked about the phases of the moon for requirement 1. See the illustration there on the moon's phases and then make your own diagrams for eclipses and the moon's phases. Your counselor will want to make sure that you understand eclipses and phases, so don't just copy drawings you find in this pamphlet or in astronomy books. Your counselor will not be as impressed by your artwork as he will be by your understanding.

# Finding Your Position by the Sun

A knowledge of astronomy is of practical value to an outdoorsman like a Scout. If you know something about the movements of the sun, you always can determine north and south and you can find your latitude.

To determine a true north-south line (a meridian), pick a day when the sun is strong enough to cast a shadow. The area you choose must be in sunshine all day. Start the experiment in the morning.

Get a pole or stick 2 or 3 feet long and drive it a few inches into the ground as straight as you can. Tie a piece of string at the base of the pole. Use the string to draw on the ground a circle whose radius is the length of the pole's shadow. Now drive in a short stick at the point on the circle that the shadow just reaches.

As the sun rises the shadow will get shorter, but in the afternoon the shadow will again touch the circle. Mark that point with another small stick. The halfway point between the two sticks will be due north of the longer pole. Scratch a line from the bottom of the pole to this halfway point and you have your north-south line.

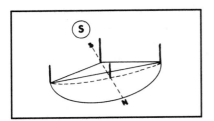

**Laying out a meridian**

## Finding the Sun's Altitude and Your Latitude

Finding your latitude is a bit more complicated. First, what is latitude? It is the distance you are from the equator, expressed in degrees. The equator is 0 degrees latitude, the North Pole is 90 degrees north latitude, and the South Pole is 90 degrees south latitude. The 48 conterminous states of the United States are located between 25 and 50 degrees north latitude. Hawaii is at about 20 degrees north latitude, and Alaska is between 54 and 70 degrees north latitude.

To find your latitude using the sun, you need to know the sun's latitude in the sky. This is called its *declination*. It is the angle the sun makes with a line extending out from the equator. A table of the sun's declinations at different times of the year is on page 39. Note that in spring and summer the declinations are north; in fall and winter they are south. This is because the sun appears to move northward or southward during the different seasons. Actually, as you know, the sun does not really move at all in relation to the Earth. Rather, the apparent movement is due to the fact that the Earth's axis is tipped at varying angles to the sun during the different seasons.

Now, at exactly noon on the next sunny day, find the altitude of the sun. This can be done with a protractor. All you do is measure the angle between the horizon and the sun. Simply measure the angle cast by the shadow of your pole.

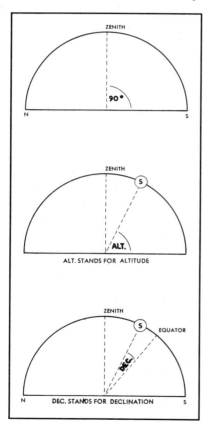

Now let's try to picture what you are going to do to figure your latitude. Imagine the sky as a huge half sphere or bowl turned directly over you. The point on that sphere directly over your head is called the *zenith*. A line from where you are standing to the zenith makes a 90-degree angle with the horizon, which is the north-south line in the drawing on the opposite page. You have used your protractor to find the altitude of the sun in the sky at noon. Subtract the altitude from the 90 degrees to get the angle from the zenith.

Now you must figure in the sun's declination—the angle it makes with the Earth's equator. On the drawing this line is marked *E* for equator. Because it is spring or summer in the drawing, the sun is high in the sky and the declination would be north. If it were fall or winter, the sun would be lower and the declination would be south.

Here is a simple formula to work out your latitude from the sun's altitude:

Latitude = 90 degrees — sun's altitude + or — declination.

If the sun's declination is north, it is plus; if the declination is south, it is minus.

Now let's try an example. Suppose you live in Quincy, Ill., and the date is May 1. You go outside at noon and find the sun's altitude is 65 degrees. You look up the sun's declination for May 1 and find it to be 15 degrees north.

Apply the formula and you have:

Latitude = 90 degrees — 65 degrees + 15 degrees = 40 degrees.

Quincy's actual latitude is just under 40 degrees, so your answer is good. Your answer will not be absolutely accurate because it is impossible to get an exact reading of the sun's altitude with a protractor and also because the figures for the sun's declination are only approximate. The table of declinations is not complete for every day of the year. Choose the date nearest the date you are working.

You also can determine your latitude using the North Star. This is a simpler method. You can make a device to determine your latitude using a yardstick, a protractor, a string, and a penny. Tape the protractor to one end of the yardstick. The penny is taped to the end of the string to act as a weight. The other end of the string is taped to the protractor so the penny swings from the center.

Sight the North Star along the yardstick and read off the angular distance of the hanging string from the 90 degree point on the protractor. This angle is equal to your latitude.

This method works since the altitude of the North Star is close to latitude all through the year.

## DECLINATIONS OF THE SUN

| Date | Decl. | | Date | Decl. | | Date | Decl. |
|------|-------|---|------|-------|---|------|-------|
| Jan. 1 | 23 S | | 14 | 9 N | | 6 | 7 N |
| 10 | 22 S | | 17 | 10 N | | 8 | 6 N |
| 16 | 21 S | | 19 | 11 N | | 11 | 5 N |
| 21 | 20 S | | 22 | 12 N | | 13 | 4 N |
| 26 | 19 S | | 25 | 13 N | | 15 | 3 N |
| 30 | 18 S | | 28 | 14 N | | 18 | 2 N |
| | | | | | | 20 | 1 N |
| Feb. 2 | 17 S | May 1 | 15 N | | 23 | 0 |
| 5 | 16 S | 5 | 16 N | | 26 | 1 S |
| 8 | 15 S | 8 | 17 N | | 29 | 2 S |
| 11 | 14 S | 12 | 18 N | | | |
| 14 | 13 S | 16 | 19 N | Oct. 1 | 3 S |
| 17 | 12 S | 21 | 20 N | | 3 | 4 S |
| 20 | 11 S | 26 | 21 N | | 6 | 5 S |
| 23 | 10 S | | | | 9 | 6 S |
| 26 | 9 S | June 1 | 22 N | | 12 | 7 S |
| 28 | 8 S | 11 | 23 N | | 15 | 8 S |
| | | | | | 17 | 9 S |
| Mar. 3 | 7 S | July 3 | 23 N | | 20 | 10 S |
| 6 | 6 S | 12 | 22 N | | 23 | 11 S |
| 9 | 5 S | 19 | 21 N | | 26 | 12 S |
| 11 | 4 S | 24 | 20 N | | 29 | 13 S |
| 13 | 3 S | 28 | 19 N | | | |
| 16 | 2 S | | | | Nov. 1 | 14 S |
| 18 | 1 S | Aug. 2 | 18 N | | 7 | 16 S |
| 21 | 0 | 6 | 17 N | | 10 | 17 S |
| 24 | 1 N | 9 | 16 N | | 14 | 18 S |
| 26 | 2 N | 12 | 15 N | | 18 | 19 S |
| 28 | 3 N | 15 | 14 N | | 22 | 20 S |
| 31 | 4 N | 18 | 13 N | | 27 | 21 S |
| | | 21 | 12 N | | | |
| Apr. 2 | 5 N | 24 | 11 N | | | |
| 5 | 6 N | 27 | 10 N | Dec. 3 | 22 S |
| 8 | 7 N | 30 | 9 N | | 12 | 23 S |
| 11 | 8 N | | | | | |
| | | Sept. 3 | 8 N | Jan. 1 | 23 S |

39

# Constellations, Bright Stars, and Our Galaxy

Up to this point we have been discussing the solar system, which, as space is measured, is a tight little island. It measures about 7 billion miles—the diameter of Pluto's orbit around the sun. In the vastness of the universe this incredible distance is just a stone's throw.

Now we will step out beyond the solar system to the stars, so we must begin thinking of distances so immense that we need a special measure to talk about them. That measure is the light-year. It is the distance that a ray of light, moving at 186,000 miles a second, travels in one year. The distance is almost 6 trillion (6,000,000,000,000) miles.

The nearest visible star (except for our sun) is 4.3 light-years away—about 26 trillion miles. To say the same thing another way, the light we are now receiving from that star left it more than 4 years ago. Beyond that "nearby" star are billions upon billions of other stars, many of them billions of light-years away from us.

If your mind reels at the idea of such huge distances, don't worry. Astronomers can't really comprehend such vastness either because, although they can talk in terms of such figures, the human mind has trouble grasping their real meaning.

Now let's begin our blast-off out of the solar system with a look at the constellations.

## The Constellations

Probably you know the names of a few constellations already from study in school. If so, you know that constellations are groupings of stars as they are seen from Earth. Most of the major ones were named by the Greeks, although the scientific names used now are in Latin. They were named for Greek heroes, gods, and everyday objects. There

**The Pleiades**

is Orion, the Hunter; Hercules; Scorpius, the Scorpion; Aquarius, the Water Bearer; and so forth.

As a beginning astronomer you should be aware that the stars in a constellation are not really groups at all—except as seen from Earth. Usually, the stars that make up a constellation are at very great distances from one another.

For example, in the handle of the Big Dipper the star at the end (Alkaid) is 210 light-years from us. The middle star (Mizar) is 88 light-years away, and the one closest to the bowl (Alioth) is 66 light-years from us. We are much closer to Alioth than it is to Alkaid.

If it is true that constellations are made of stars that are not really associated with each other, why do we bother about naming and learning them? The answer is simple: Constellations mark off various regions of the sky and therefore give us "addresses" in the heavens. If you were to see a comet in the constellation Orion, its location is instantly placed for another astronomer.

Astronomers have named 88 constellations. many are faint and need not concern us here. Of course, we in the United States never see some bright constellations, such as the Southern Cross, because they are visible only from places much farther south than American Scouts live. The constellations we will identify are the major ones visible in the United States. A star chart will be a great help in locating constellations as you scan the skies. Check with your counselor or library.

# CIRCUMPOLAR CONSTELLATIONS

**(Visible in All Seasons Because They Seem To Circle the North Pole)**

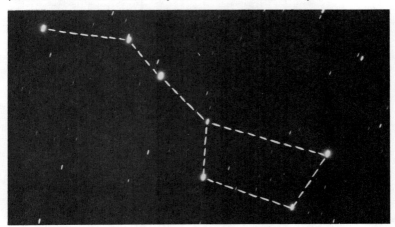

**Big Dipper (Ursa Major: The Great Bear).** You should recognize this one right away. The Big Dipper is not the whole constellation of Ursa Major; the Great Bear also includes several stars below and to the side of the Dipper's bowl. These, however, are not very conspicuous. The two stars that form the outside of the Dipper's bowl can be used to find Polaris, the North Star, since they point right at it at all times. Because the Big Dipper is circumpolar, it seems to circle the North Pole of the Earth each day—although we only see it at night. We say it "seems" to circle because we know that what is really happening is that the Earth is rotating and that we are the ones who are actually circling. In a 6-hour period the Big Dipper will turn a quarter of the way round a circle centered on the North Star.

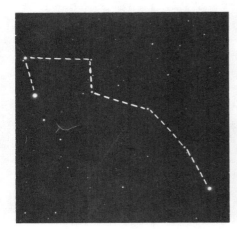

**Little Dipper (Ursa Minor: The Little Bear).** This is another constellation that should be familiar to you. You know that the North Star is the end of its handle. So if you can find the Big Dipper, you can also find the Little Dipper. It is not quite so outstanding as the Big Dipper because most of its stars are fainter.

43

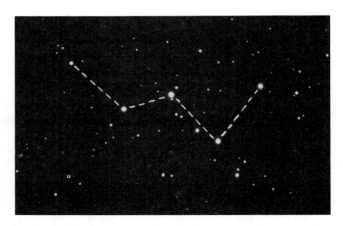

**Cassiopeia.** This constellation's five chief stars make a flattened "M" in winter and "W" in summer. It may be found by sighting from the middle star in the Big Dipper's handle through Polaris, the North Star. This line will carry your eye to Cassiopeia.

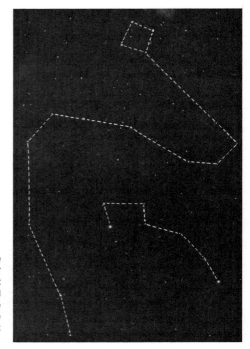

**Draco (The Dragon).** This is the fourth of the four main circumpolar constellations. It has no really bright stars. The Dragon winds around Polaris between the Big and Little Dippers. It has more than 80 visible stars. In the early summer, it is almost directly overhead.

44

# SPRING CONSTELLATIONS

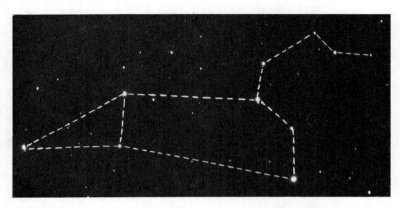

**Leo (The Lion).** You can see Leo from January through June. Find it by following with your eye a line extending downward from either side of the Big Dipper's bowl. If you extend the line from the inside of the bowl, it will carry your eye to Regulus, the brightest star in Leo.

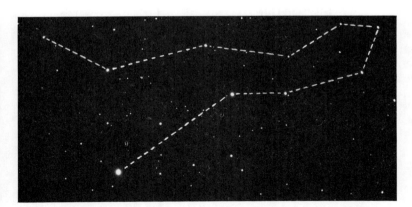

**Virgo (The Virgin).** This is one of the largest of all constellations. It may be seen from April through July. You can find it by sweeping your eye out from the handle of the Big Dipper and continuing through Arcturus until you come to another bright star. This one is Spica, which marks the bottom of Virgo's "Y" shape.

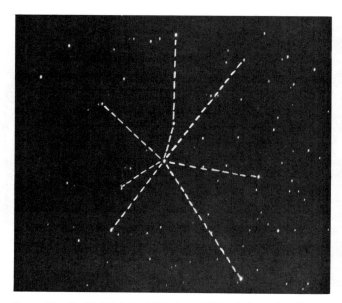

**Cancer (The Crab).** This is the faintest constellation in the zodiac. You can find it on a clear night by looking on a line between Leo's star Regulus and the Gemini constellation (see page 53). Cancer is visible from January to May.

## SUMMER CONSTELLATIONS

**Lyra (The Lyre).** This is a tight little constellation dominated by a bright star called Vega. Lyra can be seen from May through November. In midsummer it is almost directly overhead. If you look straight up in summer and see a very bright blue-white star, it must be Vega. Only two other stars that we easily can see—Sirius and Arcturus—are brighter.

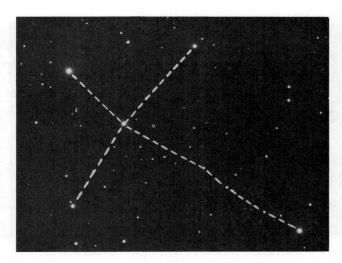

**Cygnus (The Swan).** This constellation often is called the Northern Cross. It is a beautiful sight and is visible from June through November. Look for it a bit east of the star Vega in Lyra. Its leading star, Deneb, is almost as bright as Vega.

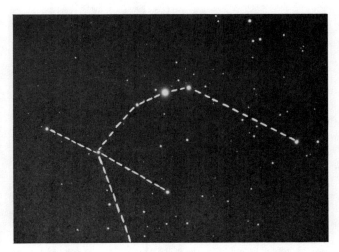

**Aquila (The Eagle).** Aquila has a roughly triangular outline and may be seen in the summer sky somewhat to the southeast. It is dominated by Altair, the third of the bright summer stars. (The others are Vega and Deneb.)

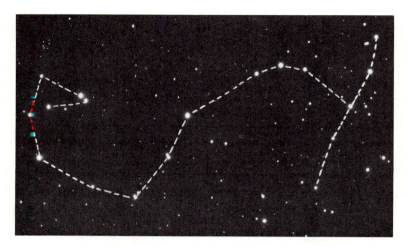

**Scorpius (The Scorpion).** Low in the southern summer sky is Scorpius, which, unlike most constellations, really looks something like its namesake. Its brightest spot is Antares, a reddish supergiant star near the scorpion's head. If you live in the North, it may be obscured by haze because it is so close to the horizon. Look for it in July and August.

**Sagittarius (The Archer).** Just east of Scorpius lies Sagittarius, which also may be difficult to find if you live in the North. It may be seen in July and August. Although it has several bright stars, none is as outstanding as Antares in Scorpius.

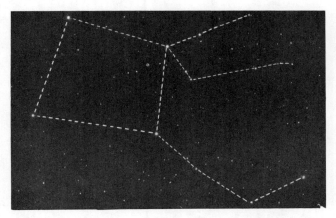

**Pegasus (The Flying Horse).** This constellation's main outline is a large square. It shares one of its stars with the constellation Andromeda. It will be found east and south of Cygnus. Pegasus can be seen best from August to October.

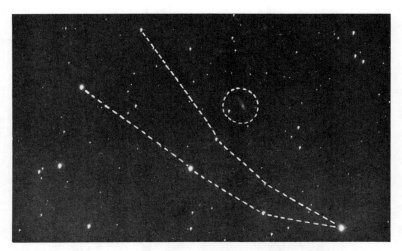

**Andromeda.** The brightest star in Andromeda is shared with the constellation Pegasus. Andromeda's chief stars form an irregular line running eastward from the square of Pegasus. It can be seen from September to January. Andromeda is noteworthy because within its borders an observer can see, on a clear night, the most distant object that is visible without a telescope (circled above). It is another galaxy, more than 2 million light-years away from us. Even on a clear night it is merely a blur of light.

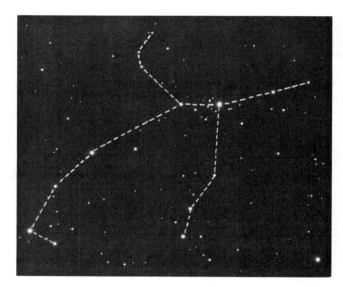

**Perseus.** This constellation is prominent in both autumn and winter skies. It is irregular in form and is to the east of Andromeda. Perseus includes a famous variable star, Algol. A variable star is one whose brightness changes from time to time.

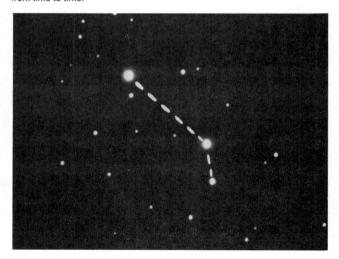

**Aries (The Ram).** Look for Aries south of the Andromeda chain from October through March. Its most prominent portion is made up of three fairly bright stars. It is one of the constellations in the zodiac.

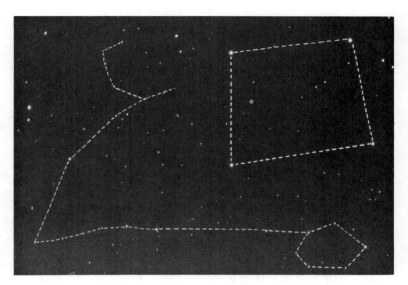

**Pisces (The Fishes).** Also found in the zodiac, this constellation is faint and hard to find. The main part of it is a string of stars below Andromeda and Pegasus.

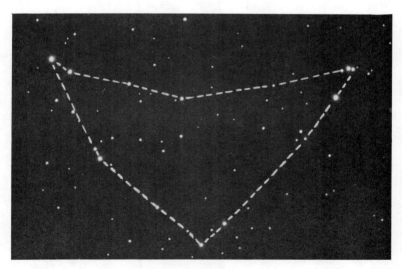

**Capricornus (The Sea Goat).** Here is still another constellation within the zodiac. Its stars are not very bright. You can look for it below Aquila in a line almost straight down from the star Altair. It is never very high up. When visibility is perfect, it can be seen from August through October.

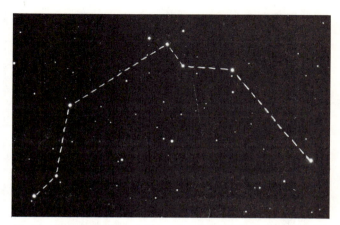

**Aquarius (The Water Bearer).** Yet another constellation in the zodiac, it is faint and formless and covers a large area of the sky. It is south of Pegasus with one end stretching above Capricornus and the other below Pisces. It can be seen with the naked eye only on very clear, dark nights from August through October.

## WINTER CONSTELLATIONS

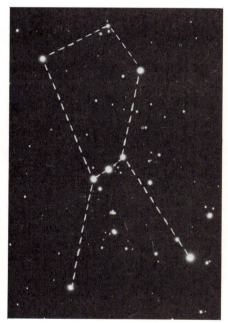

**Orion (The Hunter).** This constellation dominates the winter sky and is in many respects the most impressive of all. It has two very bright stars: Betelgeuse, a reddish-orange giant, and Rigel, a bluish-white one. Orion boasts five other bright stars as well as several less brilliant. It covers a wide area of the sky. It cannot be seen from June through August.

**Canis Major (The Great Dog).** Just southeast of Orion you will find Canis Major, which is the home of Sirius, the brightest of all stars. Because this constellation does not rise far above the horizon, the distorting effect of the Earth's atmosphere makes Sirius twinkle strongly. Canis Major can be seen from December through April.

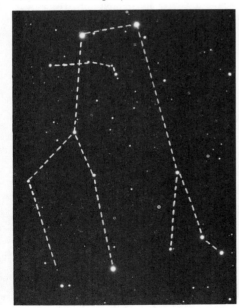

**Gemini (The Twins).** Northeast of Orion you will find a brilliant pair of stars called Castor and Pollux. They are the "heads" of Gemini. They, and their retinue of dimmer stars, can be seen from December through May. Pollux is a bright orange and its twin—which is really a multiple star—is white. Gemini is in the zodiac.

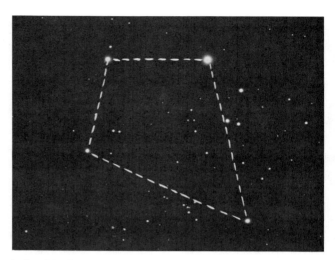

**Auriga (The Charioteer).** This handsome constellation is led by Capella, one of the six brightest stars in the sky. Like Castor, it is a double star. You can find it by sighting directly north from Orion. Auriga is most visible in the evening from November to April.

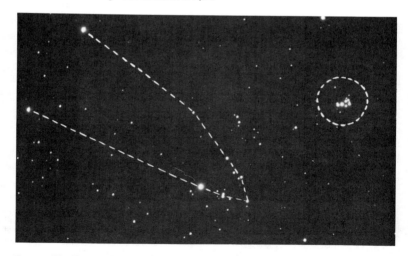

**Taurus (The Bull).** Whenever you can find Orion, you can also find Taurus because Orion's belt stars point up toward Aldebaran, a very bright orange-red star in Taurus. Within Taurus you may see, if the night is clear, two star groups—the Pleiades, a close cluster of faint stars (circled), and the Hyades, a more scattered V-shaped group. Taurus, also part of the zodiac, can be observed in the evening from November through March.

# Conspicuous Stars

The word *conspicuous* means easy to see. Therefore, when we talk about conspicuous stars, we mean those that are brightest in the sky.

Astronomers are able to measure accurately the brightness of each star. They call the brightness of a star or planet its magnitude.

There are two kinds of magnitude. First, there is absolute magnitude—that is how bright a star would be if all stars were at the same distance from the Earth. If two stars were equally bright but one was five times as far away as the other, it's obvious that the nearer star would look much brighter to us on Earth. Knowing a star's absolute magnitude is important to astronomers. However, for this requirement we will be concerned with apparent visual magnitude—that is, a star's brightness as seen from Earth.

The lower the magnitude number, the brighter the star. Stars with magnitudes of less than 0 (minus numbers) are very bright. For this requirement, let's say that if their ratings are between −1.4 and +1.5, they are first magnitude; if between 1.6 and 2.5, they are second magnitude; between 2.6 and 3.5, third magnitude, and so on. Each step up the magnitude scale is about 2½ times brighter than the next step. Stars down to sixth magnitude are visible to the naked eye in perfect viewing conditions.

For this requirement you must identify at least eight conspicuous stars, at least five of which must be of first magnitude. You can locate many such stars by consulting astronomical alamanacs or books. We already have mentioned several of them in our tour through the major constellations—Sirius, Arcturus, Vega, Capella, Betelgeuse, Rigel, Altair, Aldebaran, Antares, Spica, Pollux, and Deneb. See if you can find out what their visual magnitudes are.

# The Milky Way

Unless you live in the heart of a large city, you have probably seen on some clear, moonless night a glorious ribbon of light streaming across the dark sky. In the winter it snakes across the sky from Cassiopeia through Perseus, Auriga, Gemini, and just east of Orion before dropping over the horizon. In the summer sky it seems to travel upward from Scorpius and Sagittarius through Aquila and continues in a blaze of brightness through Cygnus back to Cassiopeia.

Its glory has been known since the first shepherds looked up from

their flocks at night and saw it as a milky pathway through the stars. It was not until the great astronomer Galileo turned his first telescope on the heavens in the 17th century that man became aware that the Milky Way is really a progression of millions of stars. With a small telescope or pair of binoculars you can make the same discovery.

The Milky Way is actually our home galaxy—a gigantic disk of 100 billion stars and vast quantities of dust and gas. It is about 100,000 light-years in diameter. Near its center it is some 15,000 light-years thick and tapers off toward the outer edge. If you took two fried eggs and placed them back to back, you would have a fair model of what the Milky Way galaxy must look like from a distance of a million light-years. Our solar system is about two-thirds of the way from the center to the outer edge of the Milky Way. The hub of this huge whirling disk lies in the direction of Sagittarius.

Our galaxy is only one of the perhaps several billion in the universe. The Milky Way appears to be an average galaxy. You can see another galaxy with the naked eye on a clear night in the fall. It will appear as a misty patch of light in the constellation of Andromeda. Astronomers call it M 31 and place it more than 2.25 million light-years from us. M 31 appears to be half again as big as our galaxy—150,000 light-years across.

Your full address in some universal post office would be Anytown, U.S.A., Planet Earth, Solar System, Milky Way Galaxy.

**Star clouds in the Milky Way**

# The Astronomer's 'Eyes'

Until the 17th century, astronomers had to depend on their naked eyes for information about the stars, moon, and planets. They were no better off in viewing the heavens than you are on a clear night at summer camp.

Early in that century the first telescope was invented, probably by a Dutch maker of eyeglasses named Hans Lippershey. His invention was seized upon by an Italian astronomer, Galileo Galilei, who was the first to build a telescope and point it toward the sky. Galileo's pioneering work with telescopes transformed astronomy because it increased man's vision hundredfold. With the improvement of telescopes through the centuries since then, our view of the heavens has deepened far beyond Galileo's wildest dreams.

On a dark, clear night you can see stars as dim as the sixth magnitude with your naked eye. With the largest telescopes available today, astronomers can collect light from stars of the 25th magnitude. The ratio in brightness between objects of the first and 25th magnitude is about 40 billion times.

There are two basic types of optical telescopes—*refracting* and *reflecting*. Let's consider them.

## Refracting Telescopes

When you hear the word telescope, you probably think of the kind called a refractor. To refract means to bend, and this is exactly what a refracting telescope does. It bends rays of light.

At its simplest, it is a long tube with lenses at each end. Light from the object you are looking at is collected by the big lens, called the *objective*, at the front of the telescope. The lens focuses the light to form an image near the other end of the long tube. The smaller lens, called the *eyepiece*, is placed so the astronomer can easily view this image.

The eyepiece is a magnifying lens which enlarges the image produced by the objective. In modern astronomical telescopes there will be two or

**The Hale telescope on Mount Palomar in California**

more lenses for both the objective and eyepiece lenses to eliminate distortions due to the nature of light and lenses. Basically, all refracting telescopes work on the simple principle shown.

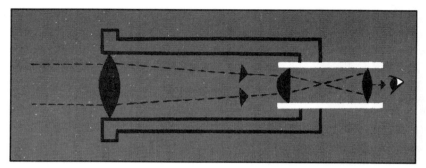

**Principle of refracting telescopes**

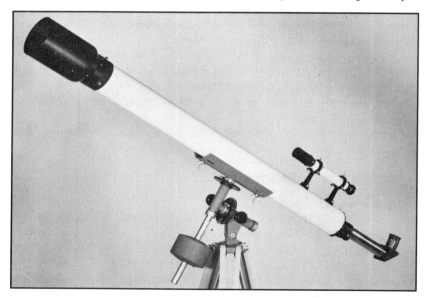

**Refracting telescope**

# Reflecting Telescopes

The reflecting telescope has no objective lens. It uses a curved mirror to focus light from space.

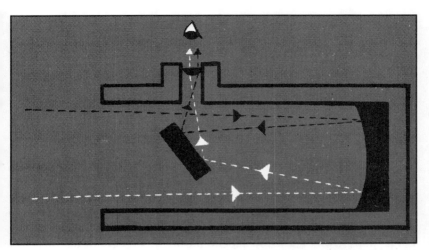

**Principle of reflecting telescopes**

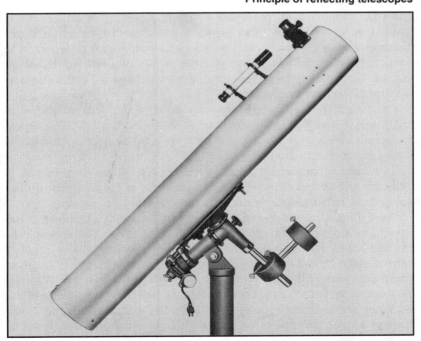

**Reflecting telescope**

Light comes into the wide opening in the front of the telescope and strikes the concave mirror at the other end. It bounces off the curved mirror surface and is reflected in a narrowing cone toward the smaller mirror just beyond the eyepiece. This mirror is set at an angle of 45 degrees to the reflected light rays, so that they are bounced up through the eyepiece lens, which magnifies them just as it does in a refracting telescope.

## Which Is Better?

Both refractors and reflectors have given invaluable service to astronomers. Each has certain advantages but, when everything is considered, the reflector is the more valuable telescope to professional astronomers. This is mainly because it can be made larger than the refractor.

The powers of a telescope depend mostly on the size of the objective lens (in a refractor) or the main mirror (in a reflector). Because a lens must be supported only around its edge, large refractor lenses are limited in size. On the other hand, a reflector can be much larger because the weight of the main mirror can be supported underneath. In theory, the mirror could be any size the astronomer wanted. However, mirrors bigger than the largest now in use would be difficult to support accurately and very expensive.

The largest refracting telescope has an objective lens 40 inches in diameter. It is at Yerkes Observatory at Williams Bay, Wis. The largest reflecting telescope is a 5.9m (236-inch) reflector near Zelenchuk-skaya in the southwestern Soviet Union.

If you are thinking about getting a telescope, consider these factors: A reflector will be less expensive than a refractor of the same size collector (lens or mirror), but a refractor is somewhat more rugged.

Don't think a telescope can have unlimited magnifying power. If you could pay any price, the greatest practical magnifying power would be no more than 50 times the diameter of the objective lens or primary mirror given in inches. Figure magnifying power as the focal length of the main element divided by the focal length of the eyepiece. You can buy eyepieces with very short focal lengths, but they are difficult to use and they will magnify imperfections in the air and the telescope. If you expect to see celestial objects as clearly and impressively as in photographs, you will be disappointed with the usual telescope you could buy.

Common binoculars (7x50) have a magnifying power of 7 and objective lenses 2 inches (50 millimeters) in diameter. A 2½- to 3-inch telescope will not give you a much better view and will cost more than five times as much.

Buy binoculars and use them to explore both earth and sky. If you like what you find in the sky, give some thought to building your own telescope. Ask your counselor or science teacher about it. All it takes is patience and care. It will take months of work to complete, but thousands of people find it fun and rewarding to build their own telescopes.

# Other Tools of the Astronomer

Telescopes are the primary tools of the modern astronomer, but he also depends on a number of other instruments. These other instruments help him to see more and to analyze what the telescope brings to him.

### Camera

Until the middle of the last century, the astronomer could not make completely accurate records of what he saw. The best he could do was make drawings of what the telescope revealed. Then, in 1850, the first astronomical photograph was made at Harvard University. This was a leap forward of great importance to astronomy. Not only could an astronomer now make accurate, permanent records of the heavens, he could "see" even more through the camera's lens than through his naked eye. The reason he could see more is that photographic film handles light differently from the human eye.

When you look at something with your naked eye, you get all the impression of light at first glance that you're ever going to get. No matter how long you look, the object you're looking at will not get brighter or more distinct.

Photographic film will continue to store up light as long as the light falls on it. This means that it will record light from objects so far distant that you cannot see them at all when you look through a telescope. The light is simply so faint that your eye will not register it. And so, most of the big observatory telescopes now are used mostly as cameras. An astronomer might look through his telescope once in a while, but most of the time it is used for taking pictures.

Because most astronomical photographs require long exposures (often several hours for galaxies millions of light-years away), the astronomer must compensate for the rotation of the earth. If he didn't, his film would show only streaks across the sky, not sharp pictures of what he was photographing. This is done with motor drives which carefully move the telescope counter to the movement of the Earth.

## Spectroscope

Ask your school science teacher whether the school lab has a prism, which is a piece of solid glass of three or more sides. If the lab has one, ask to use it. Place it in darkness and then let a sliver of sunlight or other strong light fall on it. You will find that the prism breaks up the entering sunlight into many beautiful colors. The light enters the prism as white and leaves as red, yellow, green, violet, and other shadings of color. The reason for this is that light is made up of energy waves of different lengths. When sunlight passes through the prism, these different wavelengths are bent or refracted at different angles. The longest waves we see are red. The shortest are violet. In between are orange, yellow, green, blue, and indigo.

Astronomers use a prism to learn more about the heavens with a device called a *spectroscope*. The light waves captured by a telescope from a star or planet enter the spectroscope through the tiny slit at its top. They fall upon a lens called a *collimator*, which makes them parallel. Then they pass through a prism, are focused by an objective lens, and then are seen through an eyepiece. The result is a series of colored bands called a *spectrum*. More commonly, photography replaces the eyepiece, and the instrument is called a *spectrograph*.

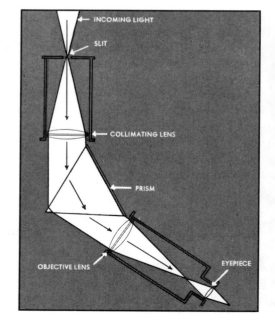

**Diagram of a spectroscope**

Because each chemical element in the gaseous condition produces a different spectral pattern, astronomers can tell a great deal about the chemical composition of a star by analyzing its spectrum. They also can take its temperature and determine whether and how fast it is moving toward or away from the Earth.

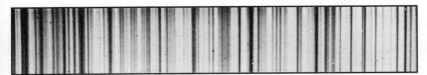

**Spectrum of a star**

### Photometer

If you're a camera shutterbug, you may have a *photometer*. It's your light meter.

A photometer is a device for measuring the intensity of light. In astronomy it is used to measure the brightness (magnitude) of a star or planet.

Photoelectric photometers achieve this result by measuring the number of electrons emitted by a sensitive chemical when the light of a star strikes it. The number of electrons released by the chemical depends on the intensity of the light.

### More Tools

Astronomers use other instruments besides the important ones mentioned. A few are the interferometer, thermocouple, coelostat, bolometer, and blink comparator. Electronic devices called *image intensifiers* are being used more and more. You can find out about them by checking books on astronomy or by asking your merit badge counselor.

### Radio Telescopes

One of the most important tools of all is the *radio telescope*. It is a relative newcomer to the astronomer's kit and has enabled him to "see" farther into the universe than is possible with optical telescopes.

Light, we have learned, is only one form of electromagnetic energy—so are cosmic rays, radio waves, infrared and ultraviolet radiation, and X rays. We can see only the waves called light, which are only a tiny part of the whole electromagnetic spectrum. Through radio astronomy we can receive and measure radio wave energy from the sun, some of the planets, galaxies, and from the dust and gases in interstellar space.

The first radio telescope was built by a young engineer named Karl Jansky in Holmdel, N.J., in 1931. With it he received radio waves from the center of our galaxy—34,000 light-years away. His work was continued by another young engineer named Grote Reber, who built the first saucerlike radio telescope in his backyard in Wheaton, Ill.

**The National Radio Astronomy Observatory's 140-foot radio telescope at Greenbank, W. Va.**

Since World War II, radio astronomy has soared in importance. Large radio telescopes are now found in every part of the world. The largest steerable radio telescope, a saucer which can be aimed in any direction, is 328 feet (100m) in diameter and is at the Max Planck Institute near Bonn, West Germany. An even larger instrument is in a natural bowl at Arecibo, Puerto Rico. It is 1,000 feet (305m) in diameter but only may be steered slightly.

These gigantic instruments are so sensitive that they can record faint radio waves that may have been traveling through space for several millions of years. They have been used to study the moon and nearby planets in the solar system by radar. Radio waves are beamed to these bodies and bounce back to be collected by radio telescopes. Analysis of the radar results has been valuable in recent years in determining the rotation periods of Venus and Mercury.

In addition, radio telescopes have helped astronomers to identify quasars, which may be the most remote objects in the universe. Through optical telescopes quasars look like average stars, but the radiation received from them by radio telescopes indicates that they are far from average. They produce energy equivalent to that of a huge galaxy, but they seem to be all in one piece. Some may be 10 billion light-years away from us and may be traveling much faster than any known celestial object. Unraveling the mystery of quasars may be the next major breakthrough for astronomy.

## Astronomy from Satellites

You probably have seen photographs of the planets taken from space probes. Galileo would have been amazed by them. Since our atmosphere blocks most of the electromagnetic energy from reaching the ground, satellite observatories can receive the whole range.

Astronomers have made discoveries in gamma rays, X rays, and ultraviolet radiation using satellites. They have found many mysteries out in deep space. The space telescope to be launched by the space shuttle in 1985 will increase the distance astronomers can "see" by more than seven times and will be able to detect objects down to the 29th magnitude.

What will be found? No one knows.

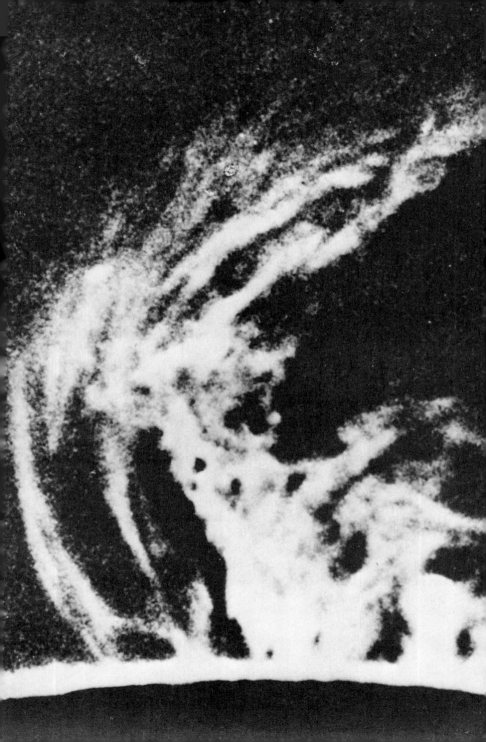

# The Sun

**Reminder:** Never look at the sun through binoculars or a telescope. You may look with your naked eye using several pieces of fully exposed and developed *black and white* film.

The sun is a star. To us it is by far the most important star because life could not exist on Earth without it. However, it is a rather run-of-the-mill star—neither very big nor very small as celestial measurements go.

Let's review a few facts about the sun:

• It is roughly 93 million miles from Earth.

• It is made up entirely of gases—more than 75 percent of it is hydrogen, which is by far the most plentiful element in the universe. The sun has at least 67 of the 92 natural elements known on Earth, and it may have all of them.

• It is 865,370 miles (1.39 million km) in diameter and weighs about 2 octillion tons. It includes 99.86 percent of the mass of the whole solar system.

• Its surface temperature is 10,300°F. At its core, it may be as hot as 30 million degrees F.

## The Solar Furnace

We sometimes speak of a "burning sun," but the sun does not really burn in the usual sense. If it were made of coal and were burning hard enough to send out as much heat and light as it does, it would burn up in a relatively short time—maybe 5,000 years or so. But our sun has actually been shining 5 *billion* years and it appears to be good for another 5 billion.

The sun produces energy by nuclear fusion. It is somewhat similar to a huge hydrogen bomb that is under control and that is going off constantly. Every second, the sun is transforming 564 million tons of its hydrogen into 560 millions tons of helium. The 4 million tons of matter that is lost in this fusion process is changed into the energy the sun radiates into the solar system.

**A solar prominence**

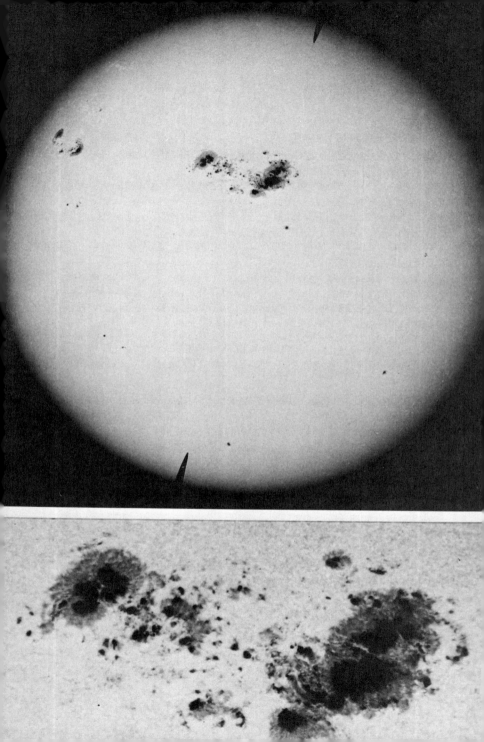

# The Sun and Our Weather

Luckily, we don't get much of this outpouring of energy. If we did, the Earth would be fried to a crisp in short order. Only one part in two billion of the sun's energy reaches the Earth. This tiny fraction is enough to sustain life. It is also enough to make our weather what it is.

If it's hot or cold, blame the sun. If it rains, blame the sun. And if the wind blows too hard, you can blame the sun for that, too. The sun's energy sucks up moisture from the earth by a process called *evaporation* to form clouds that bring our rain and snow. It heats the atmosphere, and when one air mass becomes hotter than another, winds begin. The cooler air mass, which is denser and heavier than warm air, will move to replace it, and this is the wind.

## Sunspots and Their Effects

Sunspots are dark blotches on the sun's surface. From Earth they look small, but some of them may be 50,000 to 100,000 miles across. They appear to be caused by disturbances in the sun's strong magnetic field.

They look dark because a sunspot area is cooler than the rest of the sun—about 2,000 degrees cooler. Their appearance can't be predicted with any accuracy, but on the average about 11 years passes between times of minimum sunspot activity. Some spots can be found by astronomers almost all the time.

Sunspots, or other phenomena that seem to be associated with them, have a considerable effect on Earth. Tremendous flares shooting out from sunspots send minute particles of electrically charged matter hurdling toward the Earth. When these strike our atmosphere, they cause magnetic storms. When these occur, radio-telephone communication is blacked out, radios don't work well, television sets are affected, and magnetic currents may even interfere with the ordinary telegraph system.

The solar turbulence that is associated with sunspots and flares also is responsible for the aurora borealis (northern lights). All these effects are caused by the radiation from the sun's eruptions that strikes the upper atmosphere of the Earth.

**A large and complex sunspot group**

# The Stars

We have already noted that our sun is a star. We can say with considerable confidence that all other stars are similar to the sun. There are great differences in size, chemical composition, temperature, color, and brightness—but it's all in the family. The sun is a cousin to them all.

In size, stars vary from some not much bigger than the Earth to supergiants more than 1.000 times as big as the sun. In surface temperatures, stars range from about 3,000°F. to 80,000°F. They come in all colors with red, orange, yellow, white, and blue most common.

As we view them from the Earth, stars appear to twinkle, but in fact they do not. The twinkling effect is caused by the distortation of their light as it passes through Earth's atmosphere.

More than half the stars we see are really doubles or multiples. In some cases, two stars orbit each other and are called *binaries*. In others, the "double" stars may actually be at a great distance from each other, but they are so close together in our line of sight that we see both stars as a single point of light.

## The Meaning of Color in Stars

The color of a star is evidence of two things: its temperature and the elements it contains. Generally, blue and white stars are the hottest; yellow stars, like the sun, are in the middle range of temperature; and orange and red stars are coolest.

For requirement 7b you are asked to identify one red, one blue, and one yellow star and explain the meaning of their colors. Since most stars show their true colors only through a telescope, we will mention the names of stars in each classification. But it's up to you to find them. (Tip: Put your binoculars or telescope slightly out of focus to see a star's color better.)

### Blue and Blue-White Stars

These are among the hottest of all stars. Through a telescope they appear either blue-white or very blue. Examples are Rigel in Orion, Spica in Virgo, and Sirius in Canis Majoris.

## Yellow Stars

These are a good deal cooler than the blue stars. The sun is one, and so is Capella in Auriga.

## Red Stars

These are the coolest of the main classifications of stars. They include Betelgeuse in Orion and Antares in Scorpius. Somewhat warmer than the red stars are those which appear orange-red, like Aldebaran in Taurus and Arcturus in Boötes.

It stands to reason that the hotter the star, the brighter it ought to be. So you may have asked, "How come we can see Betelgeuse so well if it's so cool?" The reason is that a star's visual magnitude depends on three things: (1) its temperature, (2) its size (these two make its true brightness), and (3) its distance from us.

Betelgeuse is only half the temperature of the sun, but it is a supergiant 500 times bigger in diameter. Because it has so much more surface area, it is equal to 17,000 suns in brightness. Therefore, even though it is 520 light-years away, Betelgeuse is one of the brightest stars in our sky.

# Viewing
# the Heavens

Whichever of the optional requirements you choose to do, you're sure to have a lot of fun. You're also sure to learn plenty about the skies and the science of astronomy.

If you would like to visit a planetarium or observatory, check with your merit badge counselor to find out if there is one near you. Many large cities and some colleges and universities have planetariums or observatories which can be visited.

At a planetarium, you will see moving models of the solar system, stars, and constellations as well as astronomical charts and instruments. At an observatory you will see astronomers at work, and you also will see how a large telelscope is operated.

After your visit be sure to write a report on what you saw for your counselor. Make it at least 500 words, but don't think you have to cover the whole field of astronomy. Your counselor will be more interested in the new things you learned than in what you knew before your visit.

If there is not a planetarium or observatory near your home, you will have to complete requirement 8 by spending 3 hours outdoors observing the night sky. If you don't have binoculars or a telescope, chances are that you can borrow them from another Scout. Otherwise, check with your counselor.

Choose a clear night and take a star chart outdoors with you so you can be sure of what it is you observe. Go as far away from lights as possible. A clearing in the woods is ideal. If you live in an apartment house, try the roof. The important thing is to be far away from artificial lighting so that the sky looks very dark. To see your star chart, you'll need a flashlight. Cover its lens with red cellophane so that the light's glare won't dim your night vision.

What should you look for? Anything that interests you. You might observe a few particular stars, identify them on your chart, and check them off. Determine their color and their brightness compared with nearby stars. Are some of them double? Look at hazy areas. Are they clouds of gas or stars? Locate them on the chart.

You might spend some time looking at the moon and perhaps make a drawing of what you see. You might concentrate on a planet and follow its progress during the 3 hours. You might see some of the moons of Jupiter. Notice how they change position in 3 hours. You might see a meteor or shooting star. If so, record the time and its position.

Observation of the moon's phases can be of practical value in planning outdoor activities at night. Shortly after new moon, the moon is seen in the western sky for increasing intervals each night as it approaches the full moon. This period covers the *waxing* phases of the moon.

At full moon, the moon is close to the eastern horizon at sunset and remains in the sky through most of the night.

During the *waning* period, after full moon, moonrise occurs from near midnight until dawn. Consequently, there is no moon in the sky during the early hours of darkness. For this reason, however, the best time of month for observing the heavens above is when the moon is in its *waning* phases. . . especially when the days are long. The darker the sky, the better the viewing.

After your observation period, write a report of 500 words or more for your counselor. Put down what you saw, not what you remember from your reading. If you have made sketches of the moon or planets, include them. Your report does not have to be elaborate, but it should tell your counselor about those things you found out by actually observing the heavens.

**A major planetarium projector**

# Viewing the Sun by Projection

Never look at the sun directly through a telescope or binoculars. Instead, project its image on a piece of white cardboard held near the eyepiece. The drawing shows the method.

To focus the sun's image on the cardboard, simply move the cardboard back and forth until it is sharp and clear. If the eyepiece gets warm, point the instrument away for a while. It will get warm if you cannot project the entire face of the sun on the screen. It also will get warm if it's dirty.

Here are some other projects you might try: Draw the positions of the sunspots on a paper held on the cardboard screen. Try to locate north and south, east and west. (Ask your counselor how, or watch how the image drifts as the sun moves westward in the sky.)

Draw the sunspots' position every day at noon for 4 weeks. What have you learned that Galileo discovered?

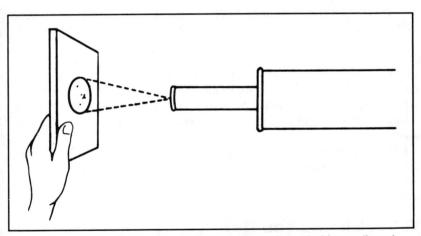

**Project the sun's image through a telescope onto a piece of white cardboard.**

# Careers in Astronomy

Now that you have traveled this far along the road to the Astronomy merit badge, you may have developed an interest in this science. Do you think you'd like to be a pro? If so, you can look forward to an exciting career.

Astronomers study the most basic questions about the physical universe. Every new discovery is one more thrilling step along the pathway of knowledge. Some astronomers are full-time research scientists at observatories, which may be operated by a college or university, the government, or a private institution. Others combine research and teaching at the college level.

In recent years astronomers have been employed by industry where their special skills are valuable, particularly in the design and use of space vehicles.

A few astronomers are employed by major planetariums to improve the public's scientific awareness and understanding. All of these astronomers must be highly qualified. In most cases, they must have a Ph.D., which requires 3 to 5 years of work beyond the college level. Their education is strong in physics and mathematics as well as in astronomy itself.

Besides astronomers, other types of skilled workers are needed in large observatories at universities. Needed are instrument makers, opticians, electronics specialists, draftsmen, computing assistants, and assistants to operate telescopes.

## What Should You Do?

If you think you would like a career in astronomy, take all the courses in mathematics that you can get in school. In addition, study physics, chemistry, and a foreign language. If you can, learn either German, French, or Russian since these are the most important foreign languages in astronomy.

When you go on to college, you probably will major in physics. Your school guidance counselor will be able to give you a list of colleges and universities that have good courses in astronomy.

# Books About Astronomy

Adler, Irving. *The Stars: Decoding Their Messages.* Rev. ed. Crowell, 1980.

Alter, Dinsmore; Clemishaw, Clarence H.; and Phillips, John G. *Pictorial Astronomy.* Rev. ed. Crowell, 1974.

Alter, Dinsmore. *Pictorial Guide to the Moon.* 3d ed. Crowell, 1979.

Berger, Melvin. *Comets, Meteors and Asteroids.* Putnam, 1981.

Berger, Melvin. *Planets, Stars and Galaxies.* Putnam, 1978.

Branley, Franklyn M. *Black Holes, White Dwarfs and Super Giants.* Harper & Row, 1976.

Dietz, David. *Stars and the Universe.* Random House, 1968.

Gallant, Roy A. *Exploring the Planets.* Doubleday, 1967.

Gallant, Roy A. *Exploring the Universe.* Rev. ed. Doubleday, 1968.

Johnson, Gaylord and Adler, Irving. *Discover the Stars: A Beginner's Guide to Astronomy.* Rev. ed. Arco, 1965.

Joseph, Joseph M. and Lippincott, Sarah H. *Point to the Stars.* 2d ed. McGraw-Hill, 1977.

Lauber, Patricia. *The Look-It-Up Book of Stars and Planets.* Random House, 1967.

Menzel, Donald H. *A Field Guide to the Stars and Planets.* Houghton Mifflin, 1975.

Moore, Patrick. *Comets: An Illustrated Introduction.* Scribner, 1978.

Muirden, James. *Astronomy with Binoculars.* Crowell, 1979.

Nicholson, Iain. *Road to the Stars.* Morrow, 1978.

Nicholson, Iain. *Simple Astronomy.* Scribner, 1974.

Nourse, Alan E. *The Giant Planets.* Watts, 1974.

Olcott, William T. and Mayall, R. Newton. *Field Book of the Skies.* Putnam, 1954.

Rey, Hans A. *The Stars: A New Way To See Them.* Houghton Mifflin, 1976.

Sagan, Carl. *Cosmos.* Random House, 1980.

Simon, Seymour. *Look to the Night Sky: An Introduction to Star Watching.* Viking, 1977.

Taylor, G. Jeffrey. *A Close Look at the Moon.* Dodd, 1980.

## Boy Scout Literature

*Space Exploration* merit badge pamphlet.

## Other Publications

*Astronomy*, a monthly magazine in easy-to-understand language, 411 East Mason Street, Box 92788, Milwaukee, WI 53202.

*Catalog of Astronomical Publications*, Hansen Planetarium Publications, 15 South State Street, Salt Lake City, UT 84111.

*Catalogue of Astronomical 35mm Slides*, Star-gate Planetarium Slides, Educational Materials Center, 5400 Mayflower Lane, Las Vegas, NV 89107.

*Griffith Observer*, Griffith Observatory, 2800 East Observatory Road, Los Angeles, CA 90027.

*Mercury*, the bimonthly journal of the Astronomical Society of the Pacific, 1290 24th Avenue, San Francisco, CA 94122. (The Society also publishes a pamphlet on buying a first telescope, $2, Telescope Guide Department, above address.)

*Reflector*, official publication of the Astronomical League, 7035 North Willow Wood Road, Peoria, IL 61614.

*Sky and Telescope*, monthly magazine, Sky Publishing, 40 Bay State Road, Cambridge, MA 02138.

## Acknowledgments

The Boy Scouts of America gratefully acknowledges the invaluable assistance of the staff and use of facilities of the American Museum —Hayden Planetarium in the preparation of this pamphlet. Editorial assistance was provided by Dr. Fred C. Hess, lecturer; and Dr. Kenneth L. Franklin, astronomer, who also provided assistance with this revised edition.

## Photo Credits

American Museum—Hayden Planetarium—pages 24, 40, 43–54, 72, and 76.

H. Armstrong Roberts—page 58.

Edmund Scientific Company—Pages 60 and 61.

*Experiments in Sky Watching*, illustrations adapted from originals by Helmut Wimmer (copyright ©1959), by Franklyn M. Branley, Thomas Y. Crowell Company, New York—pages 28, 32, 33, and 34.

Hale Observatories—pages 6, 18, 19, 23, 68 and 70.

Lick Observatory—pages 17, 21, 25 and 56.

McMath, Hulbert Observatory, University of Michigan—page 65.

*Naked Eye Astronomy*, illustrations are reproduced from *Naked Eye Astronomy* (copyright ©1965) by Patrick Moore. Used with permission of publisher, W. W. Norton and Company, Inc.—pages 30 and 77.

*Windows to Space*, illustrations adapted from original by John Polgreen, 1967, James Pickering, Little Brown and Company, Boston —page 64.